AS THE WORM TURNS:
NEW AND EASY METHODS FOR RAISING EARTHWORMS

By

ROY AND DIANNE FEWELL

Illustrations by
Teresa E. Adkins

Published by
Shields Publications

© COPYRIGHT 1995
© REVISED EDITION COPYRIGHT 1998

BY
ROY AND DIANNE FEWELL

All rights reserved. No part of this book may be reproduced or transmitted in any form or by any means, electronic or mechanical, including photocopying, recording, or by any information storage or retrieval system, without permission in writing from the copyright owner.

Printed in the United States of America
By
Shields Publications

ISBN # 0-914116-34-7

CONTENTS

Introduction	1
Is there a Real Market for Earthworms?	3
What Makes Redworms Different?	9
Building Beds	12
Bedding Materials	18
Stocking the Beds and General Care	23
Feeding Redworms	28
Controlling Pests	31
Coping With Hot and Cold Climates	33
Harvesting the Worm Crop	35
Packaging Options	40
Fattening Worms	44
Marketing Redworms	46
Index	54

INTRODUCTION

Redworms are very accommodating and simple to raise. Unfortunately, earthworm producers have never exchanged information or organized national groups such as other livestock producers have done. As a result, very little information has been available to those who wish to grow this very valuable commodity. A world wide shortage of earthworms exists even as new markets for them open up.

Until recently, most commercially raised earthworms have been sold as fish bait (billions of them, in fact).

Now, in the interest of ecology, earthworms are also being used to compost organic waste. Composting lessens the burdens on landfills. Research is being conducted world-wide. Gardeners and farmers alike are using earthworms to rebuild over-used soil that in the past was treated only with commercial fertilizers and pesticides. These chemicals produce plant growth but do nothing to improve the soil, so with time, the health of

the soil declines.

Some people view redworms as "miniature factories" that can create natural fertilizer from organic waste. School children are now learning about composting and some schools have set up vermicomposting (composting with the aid of earthworms) bins in the classroom to recycle cafeteria food waste. The organic fertilizer produced is used on school gardens. A few schools have utilized their ecology studies project to raise funds by selling the organic fertilizer (castings) produced to their communities.

Modern methods of growing, harvesting and selling earthworms are proving to be much more efficient than the old. You may want to raise redworms to recycle household or garden waste, and to produce organic fertilizer for your own use. You might want to fish with them or raise them to sell on a large or small scale. Few books have been written about earthworms and the useful tasks they perform for mankind. This book contains practical advice for the newcomer to this hobby and unusual but highly profitable business. It is simply a guide, not a strict set of instructions. It is an overview of some of the methods that have been proven successful. Redworms can be grown under widely varying conditions. Adapt what is practical for you in your own situation.

It is a fact that raising earthworms can be enormously rewarding from both monetary and personal standpoints. Most people will find it quite easy, even though there is a little work involved.

If you like natural things, growing redworms is for you. Raising redworms is definitely for you if you derive satisfaction from watching things grow and would like to make a little money while you're enjoying yourself!

IS THERE A REAL MARKET FOR EARTHWORM$?

You bet there is! Literally millions and millions of dollars worth of earthworms are quietly sold each year by everyday people like you and me. Did you ever wonder how these worms got to the sporting goods or gardening store to be sold (and WHO raised them)? Growing redworms is a full time, profitable business for many. For others redworms provide supplemental income. In many instances people start raising earthworms as a hobby for their personal use; for fish bait, and especially those who feel strongly about ecology and have the desire to do something positive to recycle food waste. Frequently, with a little time to develop, some realize the monetary potential and decide to grow a hobby into a full time business. Most commercial growers operate as independent business owners.

Virtually all growers, who learn to grow a quality redworm and who consistently deliver it to their customers in good condition, never have to search very far for new business. Instead, they have to make decisions as to how big they want their business to be. Word travels fast. Quality redworms are always in demand.

Earthworms are sold to a variety of markets and put to a variety of tasks. Fishing is a world-wide sport and millions and millions of redworms are bought by anglers to be used as fish bait. Fishing with live bait is always dependable and the traditional way to catch fish. Nothing is more appealing to a fish than a fat and juicy worm, a natural food for fish. Even those fishermen who use artificial lures will buy live worms as insurance for a good catch.

Redworms are a consumable product. Most successful salespeople will agree that to have consistent sales your product should be consumable. As the population of the world grows, every year there are more gardeners and anglers to buy worms. Even though new growers start growing redworms each year, the supply is never enough to satisfy the demand, especially during the peak spring season. If you don't live near a good fishing area it doesn't make much difference. As many live worms are sold by mail order than are delivered in person to the consumer. Sometimes bait and tackle stores order their live bait through the mail. Often times the buyer and grower, who may have done business with each other for years, have never met each other in person.

There is a large market for redworms to be used for so-called "breeding stock". The end consumer may be a new grower stocking new beds or an established grower restocking old ones. It could be a wholesaler who doesn't grow any redworms at all, but merely puts them in cups to sell to the local market. Many times less expensive bedrun (mixed sizes) are bought and fed in fattening beds much as cattle are fed in feed lots before going to market. Even large volume producers sometimes have so many sales that they deplete their stock and look to other growers to fill their customers' orders.

Many growers only sell locally, by the cup, to sporting goods stores, local fishermen and gardeners. Others, because of local competition, only sell through the mail. Some volume producers never sell in units of less than 10,000 worms!

You may be able to sell worms by the cup to a wholesale fish bait company in your local area. Some wholesalers, because of time constraints, raise no live bait on their own, rather, they buy worms for resale.

Advanced contracts for a whole season with shipments being made weekly are routine in this business. Finding these markets is relatively easy once production is established and you know how many worms you can supply on a regular basis.

Depending upon the competition in your local area, you may want to try route sales. Usually, local fish bait stores have regular wholesale suppliers who furnish them with live bait. Offering live bait brings in customers who purchase ice, cold drinks, food and snacks too. Bait sales often mean the difference between a mediocre or highly profitable business for many small stores. Poor service from another supplier in the past may make a local store look for a new source. In some areas the live worms are marked up as much as 100 percent from the wholesale cost. Because live bait sales mean BIG profits for them, stores will want to be sure that you can deliver a consistent supply of quality worms.

Most of the United States is well covered by live bait route salespeople. Don't think they are all men, a good many are women. Women do remarkably well at both growing and selling live worms.

Sometimes the market will be open to you because the local grower is overwhelmed with more orders than he can supply. Growers eventually retire too, with a new grower needed to fill the demand. Check out the nearby stores. Check your zoning laws, too. With a sign advertising your redworms you might be able to sell earthworms at retail prices from your home. If you are away from home during the day you could try an "honor system" box where people could deposit money into a slot to pay for a cup of worms. More details are in the chapter on selling worms.

Gardening magazines regularly feature articles on the benefits of using earthworms for composting and soil improvement. Even our national White House in Washington, DC is now beginning to compost yard and garden waste. Because earthworms are natural tillers, millions of redworms are put to work tilling and aerating compost piles and garden soils. In addition worms deposit nutrient rich castings in the root areas of plants.

Earthworms are the best friend of the farmer and soils with populations of earthworms are healthy, productive soils. Earthworm castings won't burn plants and they contain five to ten times the amount of nitrogen, phosphorus and potassium along with other nutrients than is normally present in garden soils. Most elements in worm castings are water soluble and readily available to plants.

Scientists are discovering that vermicompost is a better fertilizer than regular compost, and many countries are vermicomposting on a large scale, not just garden and kitchen waste, but industrial waste as well.

Earthworm castings are sold by the bag to nurseries and individual gardeners, and make an excellent choice as an organic plant fertilizer and soil conditioner. Thousands of gardeners buy worms and castings each year to add to gardens and compost piles. Frequently the worms are bought from seed and flower catalogs. Often the seed and flower companies do not raise redworms themselves, but mail the order to a grower who "drop ships" the redworms directly to the buyer. Look in flower and seed catalogs or the classified section of sporting and gardening magazines and you will see ads for live earthworms. Often growers advertise a few times and then, after establishing their customer base, never need to advertise again. They have ALL the customers they can supply.

The redworm does a remarkable job of recycling organic waste. With landfills brimming to capacity, con-

cerned communities all over the world are now beginning to try to reduce solid waste using vermicomposting (worm composting). It's a great solution. The community reduces solid waste (which takes up space in the land fill) and, in return, gets nutrient rich compost to use on their gardens instead of chemical fertilizer. A few enlightened city governments have started furnishing to their residents small bins to be used to compost kitchen and yard waste. Small backyard vermicomposting bins can be used by almost every household to reduce solid waste.

Home composting is good for the garden, good for our planet and saves space in the landfill. Sustainable agriculture is the goal of many researchers world wide. As more businesses and private citizens become aware of the benefits of vermicomposting, the demand for redworms to be used in composting systems may become as great as the demand for fish bait.

Worms are also sold for use as pet food, to zoos and to laboratories.

The market for live worms is ENORMOUS! and international. A huge portion of the redworms raised in the southern United States are shipped to Canada and the northern United States especially during ice fishing season. Since ice fishing is very popular the demand for fish bait worms is great.

The first thing to decide is how large you want your project or business to be and how much time you have to devote to it. One small bin will probably be large enough for home composters. The majority of growers start small with a minimum investment and learn as they grow. While you are busy learning, your redworms will be reproducing at a fantastic rate. It has been estimated that 1000 mature breeders (under PERFECT conditions) will multiply into over one million egg capsules and worms of all sizes in just one year. This is only an estimate, no grower ever has completely perfect conditions, and no one has ever been able to count every worm and egg capsule accurately. Populations can be controlled by feed and the size of the growing beds. Most growers worry about producing more worms, not fewer.

Whether you choose to keep your operation small as a hobby or grow it large, there is a DEMANDING MARKET already in place waiting for your product!

WHAT MAKES REDWORMS DIFFERENT?

There are literally thousands of types of earthworms. Only a few lend themselves to commercial culture. Redworms currently comprise the vast majority of the commercial market. Depending upon which part of the world you live in, they are known by different local names; Manure worms, English Reds, Hybrid Reds, Red Wigglers, Ozark Tiger worms, etc. For simplicity here we will simply call them redworms. Most of these worms are basically the same type of worm, (Eisenia fetida or Lumbricus rubellus). They are lively, make outstanding fish bait worms and are a valuable addition to gardens and composting systems. They can be grown in most climates.

A full grown redworm is from 2 to 4 inches long. The redworm's size is perfect for fishing with a small hook, and they are great for catching most sunfish (bream), catfish and trout.

Redworms are especially suited to bin culture because of several unique characteristics:

1. Each redworm has both male and female reproductive organs, they are hermaphroditic. An exchange of sperm between two worms occurs during mating. The egg capsules are formed from a mucus tube secreted by the clitellum which is the thick girdle near the head of the body of a mature worm. During mating the eggs are fertilized by sperm deposited within this mucus tube. The mucus tube then slips off the end of the worm and forms the egg capsule. Each worm can form an egg capsule after mating with another worm.

2. They are very fast reproducers. Given the proper conditions, a mature redworm can produce one

egg capsule every 7 to 15 days, which after three to five weeks, may hatch out an average of two to four baby worms.

3. The newly hatched worms reach breeding age in two to three months at which time the clitellum or thick band behind the head appears.

4. Redworms are very hardy and are able to tolerate wide variations in temperature and they ship well in damp peat moss or other packing material.

5. Redworms make good bait worms because they are lively and can remain alive underwater for an hour or more.

6. Redworms are productive worms. They process approximately one half or more of their own weight in food or bedding every 24 hours and quickly transform vegetable waste into nutrient rich organic compost. This characteristic makes them useful in composting projects.

7. Cups of redworms do not require refrigeration. They keep well on a shelf for a week or so at normal retail store temperature.

8. They usually will not attempt to crawl away or migrate if adequate food and water are provided for them.

There are a few other types of live fish bait sold commercially. Crickets and minnows (shiners) are available seasonally. Another worm is the "Black Wiggler", "Swamp Wiggler" or "Jumper". This type of worm is raised mostly in warm climates.

The Canadian Nightcrawler or "Flattail" is also sold as fish bait.. These worms are larger worms. They are susceptible to heat and must be refrigerated for long term storage.

A fish bait worm popular in the Southeastern U.S. is the so-called "Pink" or "Grunt" worm. This is a wild worm which has not been successfully raised commercially. Pickers gather them by hand by vibrating metal rods driven into the ground. These vibrations cause the worms to come to the surface. Some say the sound created is similar to a grunt, hence the name "Grunt worm". This worm is extremely heat and cold sensitive and keeps poorly in containers.

Other popular baits sold commercially are not worms at all, but larval stages of various insects. "Spikes" and "Maggies" are blue bottle fly larvae. "Mousees" are larvae of the crane fly. "Waxworms" are bee moth larvae. "Mealworms" are beetle larvae. Most of these baits are shipped in wheat bran and must be kept refrigerated to help keep them in the larval stage.

Gather all the information you can on earthworm farming. If you know of a local worm farm that will let you visit their farm you'll gain experience as to what works and what doesn't work. Because of the lack of useful, printed information on this subject in the past,

most successful worm growers have had their share of learning experiences. One of the advantages of raising worms is that you can get started with such a minimum of expense or experience. Although there is some work involved, it's not very much. REDWORMS GROW WHILE YOU SLEEP! It's a great home-based business or hobby.

Whatever your situation, raising redworms can be profitable and fun.

You can get started with a little effort and only a very small investment. Be sure to plan your project with a tape measure and calculator, leaving room for future expansion if needed.

BUILDING BEDS

Beds should be built and filled with bedding before buying the initial worms to stock them.

It takes very little space to raise thousands of redworms, and millions can be raised in just a few square feet of space. If you only want one small bin for individual recycling use, just follow the same directions given on a smaller scale.

Earthworms have been raised in every type of container imaginable. Old washtubs and refrigerators, 55 gallon drums cut in half, milk crates and rusty tin cans have all been pressed into service to be used as growing beds. A small earthworm farm I once visited had no bins at all! The growing bed consisted of a large pile of cow manure spread under a shady oak tree!

For successful commercial production of earthworms, beds in which to feed, water and care for the worms are necessary. Accurate records can be kept as to feed requirements and the number of worms produced by each bed using various feeds. Modern day redworm producers keep records, just as other livestock growers do. Volume growers want to be able to produce a large healthy worm in the shortest amount of time with the least amount of expense. Earthworm beds may be built either indoors or outdoors. An indoor operation needs nothing more than a weatherproof roof and walls. In all but the warmest of climates indoor beds may be a necessity.

In the deep South, outdoor beds can be built.

We feel outdoor beds are easy to work. They should be constructed under natural shade if possible. If natural shade is not available, shade cloth, which is sometimes used by plant nurseries might serve to provide protection from the heat of direct sunlight. Many growers build pole barn type structures over their worm beds.

In planning the size of the worm bed you want to build, take into consideration if, in the future, you may want to harvest the worms for commercial sale. If the answer is no, then a simple box with proper drainage will do. It can be as small a 2 feet wide, 2 feet long and 1-2 feet deep. Wood, metal or plastic will all work fine. Putting many holes in the bottom of a portable bin or leaving the bottom of an outdoor bed completely open to provide aeration and drainage is a MUST! Portable bins can be set on small blocks with a tray underneath to catch the excess water that may drain from the bin. This "manure tea" can be used to water and fertilize plants. Portable bins may also be moved indoors in harsh weather (extreme heat or cold).

To raise a few worms to fish with or to recycle yard waste, fill your bed with whatever is handy in the way of organic matter; grass clippings, chopped leaves, aged sawdust, peat moss, weeds, shredded newspaper, shredded cardboard etc. Add powdered lime and water heavily. Mix and let stand a few days and then water and mix again. The mixture may go through a heat. Don't add the worms until you are sure the heating stage has passed.

If you want to build a larger bed, or want to raise worms to sell, a bed about three feet wide works best. A bed wider than four feet means reaching across the bed is impossible. Less than three feet wide is too narrow to operate most mechanical harvesters over the bed.

Beds built of 4 by 8 by 16 inch concrete block work well, they are sturdy, maintenance free and will last for years. In moderate climates lay the block to form a wall at least 16 inches high. The joints between the blocks should be filled with mortar because the newly hatched worms are very tiny and can crawl through the smallest of holes.

Lumber may also be used to construct worm beds. Wooden beds will only last a few years set into the ground, but are less expensive to build. Those who start with wooden beds always have the option of building more permanent beds later. Some growers feel that the chemicals used to treat lumber can be toxic to worms. Untreated rough cut lumber is often used instead. Before buying expensive lumber, check to see if you have a local sawmill in your community. Heart (or "fat") pine, if available, will last almost as long as treated lumber.

Rough cut boards 2"x8"x12 feet long are easy to work with. The beds can be built as long as 100 feet with dividers being placed at intervals of about twelve feet. Nail the boards on edge with vertical framing boards at each corner. At intervals of six feet or so, a vertical framing board should be nailed on the outside to help support the wall. The framing boards should be placed on the outside of the box so that it will be easier to cover the inside seam between the boards. Two boards eight inches wide placed on edge will form a wall 16 inches high. You can also use wider boards. A side wall of 24 to 36 inches may be needed in cold climates. Building a twelve foot bed from rough cut untreated lumber can be as inexpensive as thirty dollars in some areas.

If you are building beds using wood, the seam between the boards can be covered to prevent the worms from escaping. Narrow width boards can be nailed to cover the seams of the board walls. Or, heavy grade plastic can be stapled to the inside wall to cover the seam between the boards. Roofing felt works well, too. Cover only the inside vertical walls with the plastic or roofing felt. Worms do not escape from the bottom of the beds and it must be open for drainage. Worms escape by crawling through holes in the side walls or by just crawling over the top of the wall, however, redworms normally will not attempt to migrate if conditions in their bed are properly maintained.

Placing outdoor worm beds, whether constructed of wood or concrete, into the ground below the freeze line provides protection from freezing in winter. Because the temperature of the ground a few inches deep stays relatively constant, and cooler than the surface during summer, worms grown in sunken beds can escape the heat of summer and the cold of winter.

It is believed that redworms grown in a cool place grow to be larger than those grown in a hotter environment. Sunken beds more closely imitate the worms' natural environment.

In order to take advantage of this free source of heating and cooling, we set all outdoor worm beds into the ground. To build the bed, start by digging a trench. In the Deep South, four to eight inches below the surface is deep enough. In colder climates, outdoor beds may need to be placed from 12 to 24 inches deep. Beds this deep will be a little more work, but we believe that the heating and cooling effect is worth the extra effort.

Because worm beds must be kept damp and watered frequently, good drainage in the bottom of the beds is EXTREMELY necessary! Three or four inches of gravel or sand should be placed in the bottom of the trench to provide drainage. After the trench is dug, the bed is placed into the trench.

This photo shows a concrete block bed set into the ground about eight inches, the bottom of the bed has been covered with sand for drainage. The joints have been sealed with mortar.

Wooden beds may be partly or fully pre-built, block ones must be laid directly in the trench. The bed walls may also be made of poured concrete, by building a frame and then pouring the walls.

If you have the space to build long beds with dividers, rather than several small beds, caring for the redworms will be easier. Long beds make harvesting with a mechanical harvester more efficient and are less expensive per square foot to build. If you are building a long bed with dividers, you can add onto the bed as needed using the divider as the next end wall.

After the beds are in place in the trench, and a layer of sand or gravel has been added to the bottom and the holes in the side walls have been covered; hardware cloth (mesh) can be placed across the sand or gravel to prevent digging it into the worm bedding when working the beds. The hardware mesh will also prevent moles and mice from tunneling into the bottom of the worm bed. Landscape fabric (which is used to help block weeds in garden beds) can be placed on top of the hardware cloth. Landscape fabric will help prevent roots from nearby trees and bushes from coming up in your worm beds. Landscape fabric is porous enough to allow excess water to drain, but blocks weeds well.

After the bedding and worms are in place in the bed, additional landscape fabric can be used to cover the top of the bedding. It will help keep moisture in the beds but will allow the free passage of air.

BEDDING MATERIALS

Most people are surprised to learn that dirt or soil is not used in worm bedding (or just a small amount, to aid in digestion). Redworm growing beds usually contain only organic material. The bedding should retain moisture and remain loose. Worms ingest the bedding and feed as they migrate in the beds. The bedding is food and habitat. An IMPORTANT point to remember is that worms have no teeth (they have a gizzard). A hand full of sand should be added to each five gallons of bedding or feed to provide grit for the worm's gizzard. Any material which is meant to provide nourishment must be well decayed or ground into a particle size usable by the worm. Because worms cannot survive if the bedding or feed provided is too dry, it must be kept moist.

Local availability of suitable materials will dictate what type of organic mixture you will use in your worm beds. Hardly any two growers use the same formula for bedding. Since redworms thrive in decaying organic matter such as barnyard manure, leaves, grass clippings, shredded cardboard boxes and newspapers, vegetable scraps, aged sawdust, etc., almost any vegetable matter can be used if it is well decayed or ground into a small particle size.

A good formula is one part aged manure (horse, rabbit, or cow), one part Canadian peat moss, and one part aged sawdust. Poultry manure is considered to be too concentrated to be used in worm beds. Another bedding mixture is two parts decayed sawdust, leaves, hay, etc. and one part manure. Other suitable materials are: cottonseed hulls, chopped corn or weeds, shredded newspaper and cardboard. The list is long. Almost

ANY uncontaminated organic material may be used.

Cotton processing by-products make excellent bedding if they are not contaminated by certain chemical defoliants which are often used prior to harvesting the cotton. No matter which organic matter you choose to use, large pieces should be shredded or chopped to speed up the decomposition.

Powdered (not "pellet") calcium carbonate (lime) is added to all these mixtures to obtain a near neutral pH. As the beddings age, an acid condition may result. Worms grow best when the pH is kept around neutral (7.0) A pH kit, meter, or litmus paper can be bought at most feed stores or plant nurseries. Test the pH of the bedding mixture on a regular basis. To help correct a too acid condition, sprinkle about one cup of powdered lime on a four foot by twelve foot bed, then water thoroughly. It may take several weeks to correct the condition. In extremely acid beds, you may have to remove half of the bedding and replace it with new bedding.

This photo shows a large pile of leaves and grass clippings being composted.

If you are preparing bedding using fresh manure or any fresh leaves or grass clippings etc., you should first put it in a pile to age. As it starts to decay it may go through a heat. Composting (or aging) will insure that it has passed through a heat, and that there is no danger to the worms from the heat of the bedding mixture. Most growers have compost piles waiting to be used as bedding material. Bedding material can be used as soon as it has passed through the heating stage; microorganisms, bacteria and earthworms will finish decomposing the mixture.

You can raise redworms without using manure. Peat moss, newspapers, composted grass clippings, leaves, or ground corrugated paper boxes can be substituted for the manure. Just about any well decayed organic material will do. Supplemental top feeding with commercially available worm or other feed will provide extra nourishment for the worms.

At our farm we have plenty of readily available barnyard manure. You may not be so lucky. Horse or cow manure may often be available free, and even a few rabbits produce an enormous amount of manure. If you get manure from another farm, be sure to ask about their de-worming practices. Some chemicals could damage or kill your redworms if any large quantity remains in the manure. Be cautious about adding anything new from an outside source to your worm beds. The material could be contaminated with chemicals. Try it on a few isolated worms first. Also try to avoid manure with heavy concentrations of urine. Salts in urine are bad for worms.

Looking in the classified ads of local newspapers or your State Market Bulletin could produce a source of free manure or inexpensive organic materials to be used as worm bedding. Every state in the United States publishes an Agricultural Bulletin usually on a weekly or bi-weekly basis which is free to state residents. Contact your State Department of Agriculture to obtain your free subscription.

Commercial growers should pay special attention to the particle size of the bedding they are using, because in order to gather the worms to sell they must be able to separate the worms from the bedding by sifting.

Worms love newspaper. Shredded newspaper is often used for bedding by many vermicomposters. Almost all newspaper is now printed with soy based ink, however most growers avoid the color pages, which might contain metals.

All bedding materials should be damp before adding the worms. It takes at least twenty four hours or more for the bedding mixture to absorb adequate moisture. The bedding material should be mixed and watered and allowed to sit until it is damp throughout. Don't flood worm beds. You may wash away valuable nutrients. If you can squeeze only a drop or two of water from a handful of the bedding mixture it is damp enough for the worms. Bedding should be placed into the beds six to eight inches deep to start with, leaving room for additions of new bedding and feed as needed.

At some time, the bedding in the redworm beds will need to be changed. The castings must be removed or the worms will eventually start to die from the lack of proper nutrition. Some growers don't change the bedding until the beds are almost completely filled with castings, leaving little room for additional feed or bedding. Others start with fresh bedding every year or so.

As the worms eat the feed and bedding, the beds will fill with castings. The bedding will become dark, almost black, and will be of a very fine consistency. In order to recover egg capsules and mature worms from the castings, the bedding can be sifted to remove the large worms and egg capsules. The worms should be put into beds filled with new bedding.

If you want to sift in order to harvest the castings only, use a smaller screen size.

Another method is to push the contents of the bin to one side. Prepare new bedding, put the new bedding in the empty side, and then feed and water on the new bedding side ONLY. The worms will soon realize that dinner is now being served on the other side and will move into the bedding looking for food. After a few weeks you can remove the castings and add a little new bedding to the bin. Still another method is to allow the top of the bedding to dry out and then loosen the top few inches of bedding, which will cause the worms to migrate down further into the bedding to avoid the light.

Use the nutrient rich compost on your plants and gardens to rebuild the soil. Worm castings can help restore nutrients used up by plants during growing. It's all natural and earth friendly!

STOCKING THE BEDS AND GENERAL CARE

Healthy, good quality redworms should be bought to stock the beds. Beds should be stocked with 100 to 500 worms per square foot of bed surface. There are about 900 to 1200 worms in a pound of mature redworms (called "Breeders" or "Bait size") or 1200 to 2000 or more in a pound of mixed sizes (called "Bedrun"). Which is better to use? It depends on your own personal time frame. With bedrun worms, which contain some breeder size worms, you get more worms per pound. They are less expensive to buy and seem to adapt more quickly to their new surroundings. With breeder worms you have adult worms ready to produce capsules and get off to a little quicker start.

There are several reasons common garden worms are not raised commercially. Garden worms reproduce too slowly to be of use as commercial breeders and don't adapt well to bin culture. They like to migrate and may leave your carefully prepared bedding for no apparent reason.

When the package of redworms arrives, DON'T let them sit around in the hot sun! Place the redworms on the prepared bedding and they will quickly disappear into the bedding to avoid the light. Because the bedding you have prepared for them is quite like their normal habitat, the bed will suit them just fine.

Growing a large, fat, healthy redworm is not at

all hard to do. Like most living creatures, the basic requirements are WATER, FOOD AND OXYGEN.

Adequate moisture content of the feed and bedding must be maintained to allow the worms to be able to feed, reproduce and grow to maturity. In fact, worms cannot live without moisture and will attempt to crawl away or die if allowed to dry out. A sprinkling of water once a week may be sufficient in the colder months of the year. In hot, dry weather the beds may need to be watered daily or sometimes morning and evening. For small operations a water hose with a soft sprinkler head attached is all that is needed. The top few inches of the bed will always be dryer than the bottom. Be sure, especially if you are using a portable bin, that your container drains well and that too much water is not collecting in the bottom.

On our farm we use an inexpensive landscape "misting" system which not only keeps the beds moist without flooding them but provides evaporative cooling during the hot summers in the Deep South. The mist emitters are "low water volume" types which conserve water. We find this system to be far better and less time consuming than hand watering.

We located the parts to put together our system at the local Lowes Hardware Store. A section of 1/2 inch PVC pipe was clamped to the outside of each bed. The mist emitter head screws into the PVC pipe. This places the emitter low, a few inches above the surface of the bed. This placement also keeps the emitters out of the way during feeding, turning and harvesting the worms. Interchangeable misting heads allow for different spray patterns and each tube has an adjustable flow control and shut off. This great flexibility means you can truly control the moisture in each bed for better production. Installing "misters" keeps the clutter of watering hoses to a minimum. The cost of installing these "misters" is very little. The permanent watering system for a worm bed can be put together in a few minutes. If you can't find this "misting system" locally, write to "Mister Landscaper"/Maxijet, Inc., P.O. Box 1849, Dundee, FL 33838, or call 1-800-881-6294. DON'T forget to drain any exposed pipes before freezing weather sets in! We put a valve in the end of each PVC pipe and use a small air pump to blow out all the water.

This photo is of the previously shown bed, now filled with bedding and the "misting watering system" now in place.

Another major factor which will influence the quality of the redworms you grow is sufficient feed provided in the form of well decayed organic compost, ground commercial feed or manure.

Productive redworm beds require aerated and loose bedding. Gases accumulate in the bedding from the decaying of the organic matter. The bedding will also pack from frequent watering. Turning the bedding with a pitch fork every two weeks or so will allow potentially harmful gases to escape and helps keep the bedding loose. It also provides needed oxygen in the bedding. Additionally, turning will also keep weeds and roots from nearby plants and trees from growing into the beds. Worms breathe through the skin and many growers believe that plenty of oxygen in the bedding is one of the secrets of healthy worms. Use a pitch fork when turning the bedding. A shovel might kill or damage too many worms. Avoid turning under leftover commercial feed (or remove it) which might heat up and contribute to an acid condition in the bed.

You should try to maintain a near neutral pH by using lime regularly. Lime can be added, if needed, to the feed and then watered well along with the feed. If the bedding becomes too acid the worms may attempt to crawl away or come to the top of the bedding with blisters on their skin and die. In extremely acid conditions half or more of the bedding should be removed and replaced with new bedding which will help to neutralize the acid.

Well maintained worm bins do not smell. Odors are usually caused by the use of meat and dairy products (not recommended), too much feed or water, or the lack of oxygen. If you see your redworms swarming at the very top of your bin and they don't appear to be there happily eating the feed, (don't worry, you will know the difference) they are usually trying to tell you that conditions are becoming unbearable in their home.

Redworms normally do not migrate if they are provided with a good home and will stay and reproduce wherever they are fed and watered properly. However, the tendency to migrate seems greater on rainy, starless nights. In soggy, dark beds the worms will sometimes crawl over the top of outdoor beds in search of more comfortable living conditions.

Because worms are sensitive to light and will move away from it, a low wattage light burned over the beds will keep them in the beds on rainy, dark nights. Beds that do not drain well in heavy rain especially require overhead lighting at night. Since all our beds are outdoors, security lights which automatically turn on at dusk and off at daylight provide overhead lighting for our worm beds. You might have such a light in your yard that could do double duty, as ours does.

FEEDING REDWORMS

New bedding in redworm beds provides most of the initial nourishment needed by the worms. A small amount of bedding filler material should be added occasionally. Backyard composters and small volume redworm growers can make use of vegetable and fruit waste from the kitchen by chopping them in small pieces for use as worm feed.

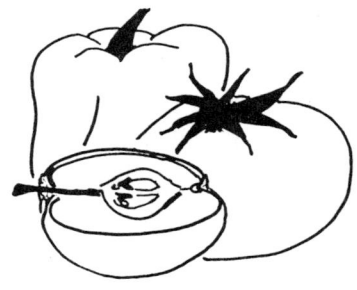

These food scraps should be buried to discourage flies and other pests. Redworms are normally surface feeders and will come to the top few inches of bedding to feed when it is dark and then migrate down into the bedding during the day.

In addition to manure and garden compost almost all commercial redworm growers supplement the earthworms' diet with ground commercial feed, much like other livestock producers supplement the diet of their livestock. If you had any doubts about this being a profitable business you should know that the giant of livestock feed, "Purina" and other feed producers such as Flint River Mills (FRM) make special feeds for worms. They contain up to twenty percent protein and are more finely ground in comparison to other livestock feed. These feeds also contain special vitamins and nu-

trients for the worms. Because of the expense of specialty feed, some growers use poultry feed. Other suitable vegetable feeds such as wheat bran, cornmeal, soybean meal, dog food, etc. are often reduced to a smaller particle size (by adding water to make a slop) for use as worm feed.

A handful of sand should be added to each bucket of feed to provide grit for the worm's gizzard.

To speed up production commercial growers sometimes re-grind supplemental feed very fine, almost to a powder, which is easier for both the mature and newly hatched worms to eat. Finely ground feed is used to produce a "fatter" worm in a shorter time. Given enough time bacteria and micro-organisms normally present in the bedding will break down the feed into a small enough particle size the worms can ingest.

Supplemental feeds are spread thinly in streaks on top of the bed every few days as needed and watered to keep the feed moist. Turning them under might cause an acid condition to develop. It is recommended to check the pH of the bedding regularly. If the pH is too low, lime can be mixed with the feed before feeding and watering. Wait until all the old feed is consumed or remove the old feed if necessary before feeding again.

Manure is sometimes used as a supplemental feed. Manure from horses, cows and rabbits (all of which are plant eating animals) is basically partially digested finely ground grass, hay and grains.

Aged manure is often used as an ingredient for bedding filler. Fresh barnyard manure (horse, cow or rabbit) can be top fed as feed. To top feed with fresh manure, spread the damp manure about an inch thick on top of the bed in streaks (it should not completely cover the bed), and sprinkle on a little lime. The lime will help control what little acid and odor that might be present.

Fresh manure will heat up as it decomposes. By placing the manure in streaks the worms have room to back away from the heat if it gets too warm. After spreading the manure and lime on top, water it well. The worms cannot eat if the feed is too dry.

Since redworms are normally top feeders, some growers cover the feeding area. There is a divided opinion on this practice. On the one hand, covering the feeding area will create a dark environment, help retain moisture and help induce the worms to feed.

On the other hand, covers can sometimes encourage unwanted insects and pests. Working the beds is a little more trouble because you first must remove the covers. If you choose to use covers they should be changed before they start to decay and pests nest in them. They should cover the surface of the feeding area only, leaving the sides of the bed and walls uncovered. Hay or straw placed on top of the bed can be used to retain moisture, but will make harvesting more difficult if it is mixed into the bedding. Landscape fabric is inexpensive and allows air and water to flow through. Other materials that might be used for covers are burlap bags, corrugated cardboard, newspapers, etc. Covers should consist of material that water and air can circulate through.

CONTROLLING PESTS

You can control most of the natural enemies of redworms by following a regular schedule of feeding, turning, watering and the use of lime to prevent an acid bed condition. Tidy housekeeping around the redworm production beds will also help. Keep the grass and weeds cut, you can compost the grass clippings and leaves for use as bedding or feed.

Some of the natural enemies of redworms are mice and rats, toads and frogs, birds, moles and shrews, gophers, and (in the Southern United States) armadillos. These pests can be controlled by not allowing trash or debris to accumulate near the beds and the use of traps, natural organic deterrents or poisons placed on the outside of the beds. Birds can be discouraged by traditional methods such as scarecrows or covering the top of the bedding.

Some insect pests to be concerned about are ants, centipedes, springtails and mites. A small amount of insecticide sprinkled around the outside of the beds will help to control these pests. All these pests will consume the feed intended for the redworms. Some believe a very light spraying on top of the bed with an insect spray (such as Black Flag) will interrupt the insects' natural life cycle. You may not kill the adult insect, but eventually the population will decline because the eggs or larva may be damaged by the insecticide. If you choose to spray always spray with caution using a mask and protective clothing. Spray very lightly, because a heavy concentration of insecticide in the bedding may damage the worms, or worm egg capsules. Anything that is poisonous to humans may be poisonous to earthworms also.

The regular use of powdered limestone (calcium carbonate) to keep the beds near neutral instead of too acid is probably the best preventative measure you can take. Most insects prefer acid conditions, so by maintaining a neutral pH in the bedding, many of the insect pests that can invade worm beds may never become a problem. Powdered lime comes in many forms and most have been used without problems; however, hydrated lime or builders lime is slightly caustic and is usually not recommended for use in worm beds. Quite a few experienced growers feel that the regular use of lime helps them produce healthy worms and will cure most of the ailments that can affect redworms.

A few good cats can keep your worm farm free of unwanted pests such as moles, shrews and mice. With cats around birds will be cautious about digging in your worm beds too.

Another factor involved in healthy beds is moisture. Too much moisture may cause "soured feed" and high mite populations. Good drainage is necessary to prevent soggy beds. Beds kept too dry may favor ants and prevent the redworms from feeding aggressively. Water according to prevailing temperatures and local weather conditions and check the moisture content of the bedding often.

Mold will sometimes grow on the surface of worm beds, but normally is not a problem for the production of redworms. Exposing the beds to strong light occasionally will usually destroy the mold.

COPING WITH HOT AND COLD CLIMATES

Redworms grow best in moderate temperatures (from 40° to 90° F). By building the beds below the surface of the ground, you can take advantage of the almost constant temperature of the ground just inches below the surface. In hot, dry conditions, sprinkling the bed with water will provide cooling in the form of evaporation. In extremely hot weather the beds may need to be sprinkled in the morning and evening, or if you have installed a "misting" system, you can let it mist the beds in the hotter part of the day. Shade, whether natural or artificial, is helpful to protect the beds from direct, hot sunshine.

In cold climates where the surface of the ground sometimes freezes several inches down, your submerged worm beds will provide better protection for the worms. By burying deep into the bed, the worms can escape from freezing.

To provide winter protection and enough food for the winter, beds are first watered thoroughly. Then a trench in the middle of the bedding is dug and filled with fresh "hot" compost and feed. A thick blanket of wet fresh "hot" compost is banked on top of the bed before freezing temperatures set in. On top of this "blanket" of decomposing organic matter and surrounding the bed, bales of hay or straw are sometimes used to provide additional insulation. As all this organic matter slowly decomposes it provides feed and heat for the worms. They still have room to move away from the heat if it gets too hot. It is important not to break the seal of this blanket while freezing temperatures are still a possibility. After the beds warm in the spring they are given a little new bedding and a good feeding of fresh commercial feed, compost or manure.

Redworms in beds located in unheated basements can usually survive the winter if they are banked with hay or straw. Beds inside outbuildings may need to be banked or heated with whatever means is economical in your area. Provide venting, because fumes from gas heaters or wood stoves can kill the worms.

HARVESTING THE WORM CROP

In the past, redworms were traditionally harvested by hand; that is by digging through the bedding and picking out the mature worms for sale. This method remains as one way to get the worms to market. However, it is very time consuming and extremely labor intensive.

Harvesting the worms by sifting the bedding is more efficient and can be done both manually or with a motorized mechanical shaker.

If you are growing redworms to use for composting yard and kitchen waste, you can add some of them along with the castings to your gardens. In the garden worms feed on organic mulches near the surface and then deposit nutrient rich castings below the surface close to plant roots. They don't eat your living plants, so be sure they have food in the garden by providing mulch. Don't forget to water them either. Redworms can be added to outdoor compost piles to speed up the action. Add the worms after the compost pile has gone through a heat and the heat of the compost pile is not enough to burn the worms.

Or, you can just go fishing with them and put the largest ones on a fishing hook. Worms are a natural food for fish and redworms make OUTSTANDING fish bait worms.

Before harvesting the worms let the bedding dry slightly and then feed and water a few days before you harvest. The old saying "The early bird gets the worm" is true! You will find the greatest concentration of worms feeding in the early morning in the top few inches of the bedding.

For small operations, a square box sifter consisting of one by six inch boards for the sides, and one fourth inch hardware cloth (mesh) making up the bottom can be used to sift the bedding. Nail thin molding strips on the edges of the bottom of the box over the wire to allow the box to slide freely on a wood frame. The frame on which to slide the box back and forth can be made from two by four inch lumber.

A simple sifter is shown here.

Build the frame long enough so that it can rest crosswise on the sides of the worm bed. It should not be quite as wide as the sifter. Place the sifter on top of the frame over the worm bed and fill it with a pitch fork full of bedding. By sliding the sifter back and forth, the bedding will drop through the hardware cloth (mesh) and along with it the smaller worms and egg capsules. The largest worms and a small amount of the bedding will remain in the sifter. Sifting over the bed means less loss of immature worms and egg capsules. After sifting, place the worms that are left in the sifter in a bucket. This method can be used quite successfully by those who are raising redworms for their own use and by most small redworm producers. The largest worms in an entire four foot by twelve foot bed can be harvested in a few hours.

The next part of the harvest will use the instincts of the redworm to finish removing the last of the bedding from the worms. Since worms react to light and will move away from it, the buckets of worms are exposed to natural or artificial bright light.

This exposure to light will cause the worms to bury into the bottom of the container, which will push the remaining bedding to the top where it can be removed. How long it takes for the worms to congregate into a mass in the bottom of the container will depend upon the brightness of the light they are exposed to.

After giving the worms enough time to make their way to the bottom of the container, the bucket or cup of worms is inverted onto a flat surface. The ball of worms can then be quickly picked up, leaving the bedding behind.

If left too long in bright light without the protection of the bedding the worms will secrete body fluids, dehydrate, and rapidly loose size. This portion of the harvest should always be done quickly to prevent possible damage to the worms. The worms should be handled gently with damp gloved hands.

Large commercial producers use a motorized mechanical harvester. Harvesting the largest worms on a regular schedule thins the population and allows more room for the young worms to grow.

This motorized shaker has different screen sizes which separate the bedrun worms from the mature worms.

Worm beds can be harvested every two to four weeks when the beds have large populations of adult worms. Rotating the harvest to different beds allows for continual harvesting, and in addition, harvesting turns and aerates the bedding in the process.

Mechanical harvesters can be purchased starting at about $1500.00. Larger ones can be as much as $3000.00. Expensive, but well worth the price if you plan to market your redworms in a big way. If you are mechanically inclined, you can build your own harvester at a fraction of the cost. Building an efficient and practical harvester is not quite as easy as it looks, and plans are strongly recommended. Building plans are available from some growers for a fee.

Mechanical harvesters are usually operated over the worm beds which helps to insure that the bedding, smaller worms and egg capsules fall back into the bed where they can grow and mature. Most mechanical harvesters consist of a slightly tilted wire mesh tube that can be slowly rotated in order to sift the bedding from the worms. Different size screens are used for harvesting bedrun or mature breeder worms.

As the screen tube is slowly rotated by a motor, larger worms in the bedding make their way to the end of the tube where they are deposited in a container. Several forks full of bedding can be sifted at one time. Motorized flat shaker type harvesters are available also. Some are not nearly as efficient as the tube design. After the largest worms have been separated from the bedding they are then placed into a bucket. Following the harvest the beds are smoothed, then fed and watered.

PACKAGING OPTIONS

Since hand counting worms is time consuming, worms are usually weighed instead. A few extra worms are added to insure fairness. In order to establish an average weight, several counts should be made to determine the weight of the number of worms you wish to place in the container. Depending upon the locality, cups of live worms may have as few as twelve or as many as one hundred worms to a cup. Check to see what is expected in your area. Nothing is more disappointing to a fisherman than not getting a true count in a cup of worms that was purchased. Bulk worms are usually sold in lots of one thousand or more. A slight overcount is normal in the industry.

After being cleaned of the bedding, weighed and placed into the container, the worms are covered with damp pre-cooled packing material. Peat moss or other packing material must be dampened with water at least 24 hours or more in advance to allow the fibers to expand. Don't soak the peat moss, dampen and stir and let it sit to absorb the water.

In extreme hot or cold weather the worms are often shipped in slightly drier peat moss which provides better insulation from temperature fluctuations. Drier peat moss takes longer to pre-cool and will warm more slowly in shipment. For short term shipping or storage additional food is not necessary. Worms packed in retail cups should last from one to two weeks before the packing material needs to be changed. Bulk shipments of worms in bags will survive for about a week also, which should be enough time for them to reach their destinations.

Gone are the days of worms sold in rusty tin

cans. Modern retailers offer live worms in attractive, vented containers.

These are some of the modern packages available. The metal object on the left is called a "worm checker", a handy little tool that allows you to dump the cup of worms to inspect them and then dump the worms and packing material back into the cup without spilling worms or packing material.

Live worm containers are manufactured by makers of containers for the food industry. Many styles are available consisting of styrofoam, wax coated paper and rigid plastic. They should have vented lids to provide the worms with air. They can be purchased plain or imprinted according to your preference and budget. Call the paper products distributor in your local area. Your first purchase will probably be small, so expect to pay more. In large quantities, cups can be as inexpensive as a few cents.

Although more expensive than styrofoam containers, our personal preference is a rigid plastic cup. These cups are made from recycled plastic, the vented lids snap together tightly and they stack well without crushing the cups on the lower level. Cups are available in 8, 12, and 16 ounce sizes. The 8 ounce size will easily hold 50 redworms. The 12 and 16 ounce sizes will accommodate 100 or more.

Feed is not usually added to the cupped worms, however, a few growers have experimented with dry or liquid nutritional supplements added along with the water used to wet packing material in an effort to extend the shelf life of cupped worms. Feeding worms in packages has always been a case of individual experimentation.

Bulk packages of worms are usually packed in bags and sold in lots of one to ten thousand. A layer of damp packing material is placed in the bottom of the bag, then the worms are added and another layer of packing material is placed on top.

"Breathable" paper bags for this purpose are sometimes used, but a new "breathable cloth" bag with tiny air holes has eliminated many of the rotting and moisture problems associated with paper bags. The seams are sewn over-lapped to keep even the smallest worms from escaping, but still allows a free passage of air through the bag.

The free passage of air encourages evaporation which has a cooling effect that helps keep the worms alive while in transit. A very light sprinkling of corn meal or other feed on the top of the packing material in bulk worm packages is not thought to be harmful. The bags are loosely filled leaving a little air space, then the top is folded over several times and securely stapled.

Up to five bags of 1000 worms each can be packaged in one corrugated box without fear of crushing the worms. Small holes can be cut in the box to allow for extra ventilation. Some growers don't use bags at all, but place the worms in a very sturdy wax lined box in damp packing material. Plenty of air space is allowed, and the box is securely sealed with tape.

All boxes containing live worms should be marked "Outside Mail" to prevent them from being put into mail sacks. They should be clearly labeled "LIVE EARTHWORMS" and "PERISHABLE", "DO NOT EXPOSE TO EXTREME HEAT OR COLD".

Because of the weight of shipping cartons, it is probably less costly for the average grower to purchase boxes from a box manufacturer nearest to his farm. Check with your local box manufacturer or Postmaster on box strength recommendations. Most cartons used for shipping live worms have an impact test of 200 to 275 pounds. A box 6x6x6 inches will hold 1000 redworms while a box 14x14x14 inches will accommodate about 5,000. Strong packing tape is used to securely seal the boxes.

Some growers have discovered that pre-cooling the packing material before adding the worms and then allowing the packages of worms to rest in a cool place before shipment results in less loss of worms.

A room maintained at a temperature of 68° to 72° F is ideal for both pre-cooling packing materials and packaged worms. Cooling to this temperature is easily accomplished with a simple room sized air conditioner. Extra insulation will help keep down energy costs. A root cellar is also ideal for this purpose if you live in an area where it is possible to build one. At least twenty-four hours in a cool room should be sufficient to allow the both the worms and packing material to cool to 68° to 72° F.

FATTENING WORMS

A cool room is also an excellent place to keep tubs in which to fatten worms before sale. At cool temperatures the redworms eat more, retain moisture and remain fatter. Mature worms (those with the "collar" or "band") are first harvested from the breeding beds.

The bedding in these tubs is shallow, only four to six inches deep, which forces the worms to be in close proximity to the feed at all times. Feed must be kept moist without allowing water to accumulate in the bottom of the tubs. The worms are usually fed straight worm feed or manure. Cover only the surface of the bedding with damp cardboard or newspaper leaving the side walls of the container uncovered. Darkness under the cover will encourage the worms to come to the top to feed. The worms will fatten and be ready for sale in about a week. The bedding must be changed every seven to ten days if the worms are to remain in the fattening tubs. Lights MUST be burned over the tubs in this instance because the worms might try to migrate because of the more dense population maintained in the fattening tubs. Light makes the best type of fence to keep worms in their homes. We keep florescent shop lights turned on twenty-four hours a day in our worm room.

Some growers don't build outdoor beds at all but conduct their entire worm business inside a building where the temperature can be maintained between 65°-75°. Bins, tubs or buckets are used to house the worms. The ever expanding population of worms is separated and split to provide enough space for the worms to grow to maturity. One method is to place about 500 breeder size worms in a small box or bucket with suitable bedding. The worms are then fed a mixture of bedding and commercial feed. The feed should be checked often and the container covered with a lid or damp newspaper which will help induce the worms to feed and mate. Lights must be burned over the containers at all times. The bedding should be changed every two to three weeks by sifting. A new bed is prepared and the old bedding which contains the breeders and egg capsules is sifted. The egg capsules and newly hatched worms are placed into a new bed where they can grow to maturity. The breeders are then placed back into the original box with fresh bedding for another round or so of breeding. Be careful about putting too many breeders in a culture box because the worms will stop producing egg capsules if the box becomes too crowded. The idea behind this practice is that worms will mate more often if they have an abundance of food and because the population is more concentrated, it is easier for them to find a suitable mate.

MARKETING REDWORMS

In order to be able to sell large quantities of redworms, you must first build up your stock, just as other livestock producers must build up theirs in the beginning. Redworm growers are at an advantage here because redworms are fast reproducers. The best advice is to start with a few beds, learn as you grow, and enjoy watching your "herd" multiply! While you are thinking about your sales plan, keep in mind that every mature worm you sell could have created generations of redworms that will be lost to you.

The decision when to begin selling your worms is one you must make based upon your own personal goals. Most commercial growers first decide how many beds they have the time to maintain. After establishing the beds, the care is minimal and quite a few beds can be maintained in just a few hours per week.

Every spring, as the ground temperature rises and the rivers and lakes warm, the fishing and gardening season gets underway with enormous worm sales. So many worms are sold that beds are often depleted and shortages are almost routine.

The most frequent mistake new growers make is to begin selling before the beds are ready and then overharvest, depleting the bed of mature worms. Time, or buying mature worms to replace some of the breeding stock that was sold are the only cures for this.

When the bedding temperature is around 60° to 70° F, production and hatching of capsules should take place rapidly. When filled with worms, a four foot by twelve foot worm bed can easily house 100,000 or more egg capsules and worms of all sizes. About one fifth of this number will probably be mature worms. Depending upon the population of each bed, as many as ten to twenty pounds of mature worms can usually be removed every month or so without decreasing the production noticeably. The immature worms and soon to hatch capsules will mature to take their place.

Dividing (or splitting the bedding) of a heavily populated bed into two beds can be used to prevent over-population in the bed and stock new beds until you are ready to harvest the worms for sale. Worms must have sufficient room to reproduce and move about.

When a bed becomes well stocked it is time to harvest or split. In order to split a bed, half of the bedding (containing half of the egg capsules and worms of all sizes) is removed and placed into a new bed. New bedding is then added to both beds to restore them to the proper depth. Splitting can usually be repeated every two to four months, as the young reach maturity and begin producing egg capsules.

The next decision to make is how you want to market the worms. Choose a simple name describing your business. Prepare sales receipts, advertising flyers, letterheads and other office supplies needed. Check to see what local and state sales licenses you may need. In addition, you should keep records of sales and all expenditures made for business purposes. Records of business miles driven should also be kept. You will be able to file an income tax return on your business. Get a copy of the tax guide for small business from the Internal Revenue Service by calling 1-800-829-1040 or from your local IRS office.

The simplest way to sell worms is directly from your farm or home, if local zoning laws permit. Advertise your business

by word of mouth to everyone you come in contact with. You'll be surprised how quickly word will spread. You should have an attractive business sign to help direct customers to you.

The worms should be packed in advance and ready for sale when your customer calls. If you don't want to be bothered by folks going fishing at 5 o'clock in the morning, or you are away from home during the day, why not sell worms by the "honor system"? Millions of newspapers are sold each day by this system. A small cooler with a jug of ice made from a plastic milk jug inside can be used to keep the cups of worms cool in hot weather. A sign showing the price and a locked slot in which to deposit money completes the set-up. Always keep the cooler stocked with cups so your customers won't have to buy their worms elsewhere.

For those planning local sales, advance preparations include: ordering containers, contacting local stores, gardening centers and fish bait wholesalers, or planning a sales route.

THIS IS A SIMPLE BUSINESS. There is no need for a slick sales campaign with a polished sales pitch, or designer clothes for that matter! Most of your customers will be independent business owners trying to make an honest living by providing good quality merchandise to their customers. They are down-to-earth practical people, just like most earthworm growers.

Be prepared to show the quality and quantity of the worms in the cup you are offering by opening the cup and dumping the contents for the buyer to inspect. Most stores that sell live fish bait have a handy little container (called a "worm checker") in which to empty the cup, inspect the worms, and then funnel the contents back into the cup, all without spilling dirt or worms.

Fish bait wholesalers normally call on a store

once or twice a week, depending upon the season, and leave a pre-determined number of cups of worms. If you decide to try route sales you will be considered a vender and should be paid for all the cups of worms you leave at the store. Modern wholesalers do not leave worms on consignment. On the next sales call, the remaining cups of worms that have not sold are each inspected and replaced if necessary with fresh cups of worms. Additional cups of worms are then left to complete the inventory level previously agreed upon. The sales receipt is written and the money collected. Most cups of worms that are picked up are not really a loss; most worms can be put back into the growing beds and will revive in a short time to be harvested and sold again in the future.

The inventory level maintained is seasonal; spring through fall levels may be double what is maintained year round. Service and simplicity is the key to local sales. Providing extra service, such as offering signs advertising live bait, marking prices on cups and maintaining an attractive and orderly fish bait display will help keep your customers.

If you find yourself picking up more than a few bad cups of worms, you may need to lower the inventory level kept at that time. Also, the store management may need information as to where to place the redworms in the store. Worms kept on top of, or close to refrigerated coolers may be dying from too much heat generated from the exhaust of the coolers in the store. Small rural grocery stores often cut off the air conditioning at night to save electricity. The heat from the exhaust of several coolers running can raise the temperature in the store by the morning, when the air conditioners are cut back on. In this case the worm cups can be put into a cooler at night for better shelf life. Direct sunlight can heat up the containers in a short period of time. You should be sure that the worms you are transporting remain cool and are not subjected to direct sunshine or excessive heat.

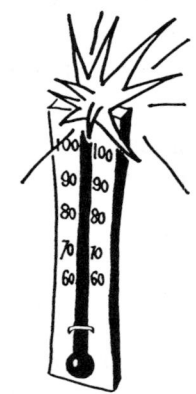

Don't over-extend yourself. This is not the place for high pressure sales. Service only as many stores as you can service in a friendly, business-like manner. In time you will make many valued friends. You're in this business not only to make money but to enjoy yourself and your customers in the process. Keep regular schedules on your route, because they will be depending upon you to furnish them with quality worms.

With the cost of gasoline and the wear and tear on delivery vehicles taken into consideration, established growers often reach out farther to the ENORMOUS national market through mail order.

Classified ads placed in special interest magazines are one of the best media for mail order earthworm sales. The largest response is probably from classified ads placed in magazines that routinely run ads for earthworms. Your local public library has books which list by categories of special interest; the address, telephone number, and circulation figures of magazines and other print media. Display advertising is sometimes a waste of money in this industry. Fishing, hunting, and gardening magazines are the place that buyers look for earthworms. You don't have to be a large volume grower to start mail order sales but you should be well established and year round production is needed.

A few ads placed in local farm and garden publications may be tried before attempting national advertising. The majority of growers start with local sales before going into mail order. Response time is important and you must send the merchandise or return the money within 30 days. Trial shipments should be made to determine if your shipments are arriving in a viable condition before placing your ads. The US Postal Service and UPS ship live earthworms routinely with timely delivery guaranteed.

Lead time should be considered when placing ads; most magazine ads are placed and paid for three months in advance of the date of publication. The season must be considered too, with most worms being sold in spring and summer months.

Newspapers (which in the past have not been considered a good advertising medium for earthworms) are usually read and discarded. Each magazine, however, is usually read by several people and is kept for months before being discarded. Orders may come in months after the ads have been run. Most established growers do both local sales and mail order selling. Local sales produce a quick cash flow and mail order provides a steady year round business.

Packages of redworms should be shipped with simple instructions on how to care for the worms once they arrive at their destination. Live shipment is guaranteed and expected in this business. After establishing their customer base, many mail order growers advertise only rarely. They have all the customers they can supply.

Keep your ads simple and truthful. The average

redworm is two to four inches long, so advertising a "Giant" might be a little silly. Read current ads from magazines and books about mail order from your local library to get an idea of how to write a classified ad that will pull a response.

Price is not the only consideration in mail order, and you should receive a fair price for your quality worms. Prices for redworms vary somewhat in different parts of the world, probably because of supply and demand. Currently, one thousand bedrun worms sells for eight to fifteen dollars. One thousand mature redworms (breeder size) sell for fifteen to thirty dollars. "Tagging" an ad by adding a letter or number to your address will tell you which ads are successful. Records of inquires will provide you with a mailing list for future advertising efforts.

Gardeners and farmers world wide are realizing the benefits of composting, and the redworm is a valuable addition to a compost project. The science of sustainable agriculture is just coming of age on a planet with increasingly limited resources. You may discover an even more useful task that redworms can perform for mankind.

Research is being conducted world wide and fellow earthworm growers will be the first to say that growing redworms is fun, productive and profitable!

If you are a little "skwormish" about worms you should investigate the facts. Vermicomposting can help solve more than just organic waste problems. Worms can produce safe, disease free organic compost equivalent to commercial fertilizers from kitchen and garden waste. What could be more natural than to return to the earth the nutrients that were taken from it to sustain the life of plants and animals. This helps complete the ecological cycle of nature.

This book is not intended to be a strict set of instructions that must be followed to the letter in order to be successful. The methods described in this book are but a few of the ways to raise earthworms. No one ever knows everything there is to know about these curious little creatures. Redworms will grow under widely varying conditions and temperatures. Common sense and a genuine love for growing things naturally should be your best guide. Many growers have raised all sorts of livestock both large and small before raising earthworms. Unlike other livestock, earthworms don't require special fencing or daily care, nor do they howl in the night or smell bad. They are simple and uncomplicated. Virtually all growers, who learn to grow quality redworms and consistently deliver them to their customers in good condition, never have to look very far for new business. Quality redworms are ALWAYS in demand.

Redworms are easy to grow, practical and EARTH FRIENDLY. You can get started today with whatever you have handy to prepare a growing bed or by planning a bigger operation if you wish. It's your choice, with no great investment needed!

The title of this book comes from a line in Shakespeare's *Hamlet*, "The smallest worm will turn being trodden on,..."

INDEX

A
Acid. see pH and Lime
Agricultural Bulletin, 20

B
Bedding materials, 18-22, see also feed
Bedrun worms, 4, 23, 52
Beds, building, 12-17
Breathable bags, 42
Breeder worms, 4, 23, 52

C
Calcium carbonate, see lime
Castings, 21-22, 35
Clitellum, 9
Composting, 1-2, 6-7, 10, 35, 52
Cooling, 15, 44-45, 48-50
Cotton by-products, 18
Cups, 40-42, 48-50
Covers, 30

D
Dehydration, 37
Drainage, 12-17, see also moisture

E
Egg capsules, 9-10, 44-46

F
Fattening worms, 44
Feed, 18-22, 25, 28-30, see also beddings
Freezing protection, 33-34

H
Harvesting worms, 35-39

L
Light, 26-27, 37, 44-45, see also harvesting and migrate
Lime, 19, 26, 29, 32, see also pH

M
Mail order, 50-52
Manure, 18-22, 28-30, 44, see beddings and feed
Marketing, 1-8, 46-53
Migrate, 10, 23, 26-27
Misting water system, 24-25, 33
Moisture, 18, 24, 32-33, see also drainage
Mold, 32

O
Oxygen, 24, 26, 41-43

P
Packaging, 40-43
Pests and pesticides, 31-32
pH, 19, 26, 29-32, see also lime

R
Redworm characteristics, 9-10
Reproduction, 8-10, 44-47
Route sales, 5, 46-52

S
Selling, see marketing worms
Shipping, 40-43, 48-52
Splitting beds, 44-47
Soil improvement, 1-2, 6-7, 52, see also composting

T
Temperatures for growth, 15-17, 33-34, 44, 46, see cooling and freezing

Turning bedding, 25, see oxygen
V
Vermicomposting, 2, 7, 52

WORM RESOURCES

CUPS
BUFORD BAIT COMPANY
P.O. Box 67
Celina, TN 38551
(615) 243-3330

ADAIR & COMPANY
Brighton, MI 48116
(313) 476-3456

WHOLESALE BAIT CO. INC.
P.O. Box 15006
Hamilton, OH 45015-0006
(513) 863-2380

WOEHRLE'S WHOLESALE BAIT
95 Church Ave.
Mount Pocono, PA 18344
1-800-227-BAIT

EARTHWORM BUYERS GUIDE
SHIELDS PUBLICATIONS
P.O. Box 669
Eagle River, WI 54521
(715) 479-4810

EARTHWORM HUSBANDRY SCHOOL
For individuals who are seriously committed to becoming a large volume bait worm producer, we offer a comprehensive, on-site, week-long training course on a limited basis.

PINEY WOODS FARM
P.O. Box 304
Montmorenci, SC 29839
(803) 641-4777

ELECTRONIC SCALE
CONSOLIDATED PLASTICS, CO.
1-800-362-1000

HARVESTING MACHINES
CYCLONE MFG. CO.
14893 El Monte Rd.
Lakeside, CA 92040
(619) 443-1698

INTERNATIONAL WORM SALES
OASIS WORM FARMS
Norma Minto
P.O. Box 15
West Union, IL 62477
(217) 279-3460
Credit Cards Accepted

MIST WATER SYSTEMS
MISTER LANDSCAPER, INC./ MAXIJET
P.O. Box 1849
Dundee, FL 33838
(941) 439-3667
1-800-881-6294

NEWSLETTERS
WORLD OF ORGANIC RESOURCE MATERIAL
W.O.R.M. Publications
P.O. Box 3458
Riverside, CA 92519-3458
(909) 681-8256

PEAT MOSS
PREMIER PRO MOSS
Fine Grind
Eastern US • 1-800-525-2553
Western US • 1-800-366-6678

SUNSHINE PEAT MOSS
1-800-964-5044

SOUTHLAND PEAT MOSS
Southern Importers
Greensboro, NC
(910) 292-4521

WORM CHECKER
SORTS-ALL
Route 1, Box 186
Cromley Lake, CA 93546
(619) 935-4227

MANY OF THESE VENDORS CARRY OTHER SUPPLIES. WRITE OR CALL FOR A CATALOG.

Other Supplies You Will Find Useful

Most of these supplies can be found at lawn and gardening, hardware, or at feed and seed stores:

- pH meter or litmus paper
- Moisture meter
- Soil thermometer
- Limestone (powdered, not pellet) or oyster flour. Limestone is used to adjust the pH of bedding materials
- Peat moss (Canadian)
- Long-handled, thin tine pitch fork (or for portable bin vermicomposters, a small garden digging fork)
- Plastic garbage cans with tight fitting lids (works great for keeping limestone and commercial feeds dry until use)
- Disposable latex gloves (can be found in the paint section of most hardware stores)
- Outdoor lights that automatically turn on at night (will keep worms home in outdoor beds on dark rainy nights)
- Small plastic cups or buckets
- Shovel and wheelbarrow
- A scale that will weigh accurately in tenths of an ounce
- Tubs with lids for storing damp peat moss
- Garden rake
- Watering can with soft spray head
- Hand sifter for harvesting or a mechanical harvester for larger operations
- Watering hoses

Specialty Items

CRICKETS
RUSSELL'S LIVE BAIT, INC.
5083 Washington Rd.
Thompson, GA 30824
(706) 595-2293

FEEDS

Contact your local livestock feed store for sources of worm feeds. Many feed companies, both national and local, produce special feeds for earthworms. They may need to special order for you. Some national companies that produce worm feed are:

FRM (FLINT RIVER MILLS)
Bainbridge, GA 31717
1-800-992-2670, Ext. 252

PURINA FEEDS

Web Sites

World wide interest in gardening and composting have created some interesting sites on the Internet. The following web site addresses should be typed in exactly as shown, as one line with no spaces.

WORLD of ORGANIC RESOURCE MATERIAL
http://www.worm-publications.com

THE COMPLETE GUIDE TO GARDEN STUFF
http://www.btw.com/garden_archive/toc.html

CITY FARMER
http://www.cityfarmer.org

GARDEN NET
http://www.olypus.net/gardens/homet.htm